**Leaping Learners
Education, LLC**

For more information and resources visit us at:
www.leapinglearnersed.com

©Sean Bulger

Published by Leaping Learners Education, LLC, Verona, NJ 07044 . All rights reserved. Reproduction in whole or inpart without written permission of the publisher is prohibited.

Every attempt has been made to credit each photo. Please contact us if there has been an error and we will resolve the issue.

Cover ; *Zakharov Evgeniy/stock.adobe.com,* Fish diver graphics © Alexey Bannykh/stock.adobe.com; Page 1, *Arnstein Rønning/stock.adobe.com*; page 2, *Andrea Izzotti/ stock.adobe.com; Petr Malyshev/ stock.adobe.com;* Page 3 *andamanse/ stock.adobe.com; laurent dambies/ stock.adobe.com; nicolasprimola/ stock.adobe.com; Melastmohican/ stock.adobe.com;* Page 4, *mirkorrosenau4/ stock.adobe.com; Vladimir Wrangel/ stock.adobe.com; fireglo/stock.adobe.com;* Page 5, *dncmrc/ stock.adobe.com;* Page 7, *davemhuntphoto/stock.adobe.com*

All design by Sean Bulger
All other photos by Sean Bulger or Royalty free from Pixabay.com

<u>Special thanks to our reading level consultant</u>
FRANCINE DEL VECCHIO, Ed.D.
Director of the Literacy Center
Professor of Education
Caldwell University

ISBN
978-1-948569-06-4

Dear Parents and Guardians,

Thank you for purchasing a *Fay Learns About* series book! After teaching students from kindergarten to second grade for more than seven years, I became frustrated by the lack of engaging books my students could read independently. To help my students engage with nonfiction topics, my wife and I decided to write nonfiction books for children. We hope to inspire young children to learn about the natural world.

Here at Leaping Learners, LLC, we have three main goals:

1. Spark young readers' curiosity about the natural world
2. Develop critical independent reading skills at an early age
3. Develop reading comprehension skills before and after reading

We hope your child enjoys learning with Fay. If you or your children are interested in a topic we have not written about yet, send us an email with your topic, and maybe your book will be next!

                              Thank you,

                              Sean Bulger, Ed.M

www.leapinglearnersed.com

Reading Suggestions:

Before reading this book, encourage your children to do a "picture walk" where they skim through the book looking at the pictures to help them think about what they already know about the topic. Thinking back about what they already know helps children to get excited about learning more facts and begin reading with some confidence.

Preview any new vocabulary words with your child. Key vocabulary words are found on the last few pages of the book. Have your child use the phrase in their own words to see if they understand the definition.

After previewing the book, encourage your child to read the book independently more than once. After they have read it, ask them specific questions related to the information in the book. Encourage them to go back and reread the section in the book to retrieve the answer in case they forgot the facts.

Finally, see if your child can complete the reading comprehension exercises at the end of the book to strengthen their understanding of the topic!

This book is best for ages 6-8
but……
Please be mindful that reading levels are a guide and vary depending on a child's skills and needs.

Fay Learns About . . . Hermit Crabs

Written by Sean and Anicia Bulger

Table Of Contents

Introduction...1

Body..3

Habitat..11

Protection..13

Diet 15

Predators..17

Fun facts..19

Glossary..20

Activities...21

Hi! My name is Fay. I love to discover and learn new things. In this book, we will learn about hermit crabs. Let's go!

Introduction

Hermit crabs are small sea animals that live on land and water. They have special **adaptations** that help them survive and stay safe in the ocean.

Tricky word: Chunk it out
Ad-ap-ta-tion

PG 2

Body

What does the body of a hermit crab look like?

Hermit crabs have a hard **exoskeleton** in front and a soft back. They have claws to grab food and protect themselves.

- Exoskeleton
- Soft Back
- Claws

PG 4

Why do hermit crabs have shells?

Because hermit crabs have a soft back, they need a hard shell to protect themselves.

PG 6

Hermit crabs do not grow their shells. Instead, they find shells in the ocean and move into new shells as they grow too big for the old ones!

PG 8

Hermit crabs have eyestalks. This is how they see when moving around.

PG 9

Eyestalk

Habitat

Hermit crabs live all over the ocean, but they can mostly be found in tide pools or in shallow water.

"Where do hermit crabs live?"

PG 12

Protection

How do hermit crabs protect themselves?

Hermit crabs protect themselves using their claws and shell. They also hide in rocks.

Rocks

Shell

Claws

PG 14

Diet

What do hermit crabs eat?

PG 15

A hermit crab's **diet** includes almost anything on the ocean floor! Some foods they like most are **algae**, **plankton,** and dead fish.

Plankton

Algae

Dead Fish

PG 16

Predators

What eats hermit crabs?

Hermit crabs have a lot of **predators**. Some predators are fish, octopus, and other crabs. When hermit crabs are close to land, birds will eat them too!

Fish

Birds

PG 17

Tricky word:
chunk it out
pred-a-tors

Octopus

PG 18

Fun facts

What are some interesting facts?

!!!!!

1. Hermit crabs live in large groups!

2. Hermit crabs can live for up to 30 years in the wild!

PG 19

Glossary

A glossary tells the reader the meaning of important words.

Adaptations - Ways that an animal changes to survive

Exoskeleton - Hard protective outer body of some animals

Diet - The food an animal eats

Algae - Plant-like organism that gets its energy from the sun

Plankton - Tiny living organisms that float and live in the ocean

Predator - Animal that eat other animals

Draw a picture of a hermit crab!

Choose 3 words from the glossary and write a sentence for each one.

1._____

_____

_____

2._____

_____

_____

3._____

_____

_____

Quiz

1. Why do hermit crabs need a shell?
a. They like the way it looks
b. It protects their soft back
c. It grows on them

2. According to the text, which animal is a hermit crab's predator?
a. Snail
b. Shark
c. Octopus

3. What is one way that hermit crabs protect themselves?
a. Hide in rocks
b. Swim away fast
c. Scare predators

4. What is the outer front of a hermit crab called?
a. Exoskeleton
b. Shell
c. Plankton

5. In which section do you learn about where hermit crabs live?
a. Body
b. Habitat
c. Diet

6. How long can hermit crabs live for?
a. 1 year
b. 20 months
c. 30 years

Answer the following questions using information from the text:

1. Where do hermit crabs live?

_____
_____
_____
_____

2. How do hermit crabs protect themselves?

_____
_____
_____
_____
_____
_____

3. Explain how a hermit crab gets its shell.

_____

_____

_____

_____

4. What does a hermit crab's body look like?

_____

_____

_____

_____

Want to learn about rainforest animals? Check out the "Matt Learns About..." Series!

- Matt Learns About... Army Ants
- Matt Learns About... Red-Eyed Tree Frogs
- Matt Learns About... Toucans
- Matt Learns About... Gorillas
- Matt Learns About... Boa Constrictors
- Matt Learns About... Tarantulas
- Matt Learns About... Anteaters
- Matt Learns About... Jaguars
- Matt Learns About... The Rainforest!

Reading comprehension strategies and activities inside!

Great for emerging readers ages 6-8

Want to learn about colors? Check out the "Clayton Teaches You About..." series!

Clayton teaches you about..... brown! (Reading Level D)	**Clayton teaches you about..... the color blue!** (Reading Level A)	**Clayton teaches you about..... green!** (Reading Level D)
Clayton teaches you about..... red! (Reading Level B)	**Clayton teaches you about..... orange!** (Reading Level C)	**Clayton teaches you about..... yellow!** (Reading Level C)

Great for early readers ages 4-6

Want to learn about Farm Animals? Check out the "Katie Teaches You About..." Series!

Great for early readers ages 4-6

Printed in Great Britain
by Amazon

24176455R00023